UNDERSTANDING TIME

What it is
and
How it Works

Edgar L. Owen

Copyright © 2016 by Edgar L. Owen

All rights reserved under International and Pan-American Copyright Conventions

First Edition, Version 1.0, September 14, 2016

Library of Congress Cataloging-in-Publication Data

Owen, Edgar L.
Understanding Time: What it is and How it Works / Edgar L. Owen – first Ed.
p. cm.
Includes biographical references.

ISBN-13: 978-1537639208 (Edgar L. Owen)

ISBN-10: 153763920X (Pbk.)

EdgarLOwen.info

CreateSpace Independent Publishing Platform

Printed in the United States of America

To my secret muse

PREFACE

This book presents a clear and easy to understand explanation of what time is and how it works based on a new understanding of science and relativity. The contents have been adapted from the author's previous work 'Universal Reality' which presents a complete Theory of Everything including the nature of time.

Universal Reality is a computational theory of reality that's completely consistent with modern science but reinterprets it in a revolutionary new way based on little recognized fundamental principles. Understanding these fundamental principles makes the nature of time clear and dispels many of the common misconceptions surrounding it.

Not only does this book reveal what time is it also reveals why it is that way. And as an added bonus understanding the nature of time automatically leads to a clear and elegant understanding of the essential points of relativity, the nature of spacetime, and the underlying reason behind the conservation of mass and energy.

Anyone interested in the nature of time or reality in general should find this book an extremely interesting and entertaining read. It convincing clarifies the nature of time and its mysteries in a refreshing new way from an entirely new perspective on the universe. It's a completely new approach to time and reality that can't be found anywhere else.

This book assumes a general knowledge of modern physics and cosmology, at least at the popular level, and some familiarity with the great perennial issues of philosophy will be helpful. But all that's really required is the desire to explore the deepest mysteries of the universe with an open mind.

Only a couple of mathematical equations are used to illustrate important concepts and where equations appear they are completely explained in English and can be skipped over with no loss of comprehension.

This book was written primarily in an effort to clarify and further develop my own understanding of time, but hopefully its publication will make it accessible to others as well and generate intelligent criticisms and suggestions for improvement. I personally believe it's the best, most accurate and complete view of time that has so far been discovered, but reality is always full of mysteries and surprises and is always the final arbiter of truth.

To the extent this book is an accurate description of time it's not something I have created, rather it's reality itself revealing itself to someone who has hopefully been able to observe and study it without projecting too much of his own personal programming and prejudices onto it. Reality is continuously revealing itself to all of us in all its awesome glory, and I believe anyone willing to observe it carefully and open-mindedly will be able to personally verify and experience the truth of much of what this book contains.

I would like to thank everyone who has helped make this book possible and encouraged me while writing it. Thanks to all of you for putting up with my unusual hermetic life style. And a special thank you to all my wild visitors, including the occasional human, and to the beauty and profundity of nature, which always inspires me with meaning and joy. Thanks to reality itself for continuously revealing itself in all its glory to those who will only look with opened eyes, and thanks most of all to my secret muse. Thank you, thank you. Thank you all!

And finally thanks to all those thinkers, scholars, scientists and visionaries throughout history without whose heroic efforts, genius and cumulative hard work this book could not have been written.

The author welcomes all comments and questions and can be contacted at Edgar@EdgarLOwen.com.

CONTENTS

UNDERSTANDING TIME 1
   THE NATURE OF TIME 1
   TIME AND RELATIVITY 3
   THE STc PRINCIPLE 4
   TIME & GRAVITATION 7
   THE VELOCITY FABRIC OF SPACETIME 9
   ACTUAL & OBSERVATIONAL TIME DILATION 12
   TWO KINDS OF TIME 13
   SOME THOUGHT EXPERIMENTS 16
   THE PRESENT MOMENT 19
   P-TIME 20
   THE ARROW OF TIME 23
   CONFIRMING A PRESENT MOMENT 24
   THE HYPERSPHERICAL UNIVERSE 25
   INFLATION & THE HUBBLE EXPANSION 28
   SEEING ALL 4 DIMENSIONS 31
   SINGULARITIES IN TIME 32
   SPACE TRAVEL 33
   TIME TRAVEL 33
   TIME REVERSAL 36
   ENTROPY & TIME 40
   ILLUSIONS OF TIME 42
   CONCLUSION 44

NOTES 46
   1. DERIVING RELATIVITY FROM THE STc Principle 46
   2. THE BLOCK TIME DELUSION 49

BIBLIOGRAPHY 52

UNDERSTANDING TIME

THE NATURE OF TIME

The central and most fundamental experience of our existence is our consciousness in a present moment through which time flows.

The present moment is simply the presence of reality, the presence of the universe. For reality to exist it must be present and its presence manifests as the present moment in which we experience our existence. Thus our experience of existing within a present moment is our direct experience of the presence of the universe within us and ourselves as a part of the universe.

This means the entire universe exists within the present moment it creates by its presence. Thus there is a universal present moment in which everything exists and there is nothing that exists outside this universal present moment because there simply is no outside. The universe and the present moment it manifests by its presence are both collocated and ubiquitous. They are all that exists. Outside the present moment of the universe there is not even nothingness. The universe in the present moment is all that exists.

Thus the past is a non-existent logical projection inferred backwards from the present. It exists only in memories and its other results in the present moment. And the future doesn't exist because it has not yet been created. Reality exists only in the present moment manifested by the presence of existence.

Note that some physicists still deny the existence of a present moment because they believe it's inconsistent with relativity but this is based on a misunderstanding of relativity as explained below. There is no doubt whatsoever that a present moment exists because it's the most fundamental and persistent of all observations both scientific and personal. The crux of scientific method is to develop theories that explain observations, never to deny them. Denying observations is the antithesis of science and the existence of the present moment is the most fundamental of all observations.

In contrast to the present moment the time that flows through the present moment, what we call clock time, is simply our experience of the continual happening of the universe. Events continually happen in the universe, processes occur and the universe evolves.

For the universe to observably exist it must continually happen. Things must change and processes must continually occur to register as experiences. Our consciousness of the flow of time depends on things changing within a flow of time because only comparison to prior states makes things observable.

Happening is what brings the universe to life and gives it reality in the present moment and sets things and processes into motion within it. Thus happening is an essential element of the existence and reality of the universe and is the source of what we call clock time. Without happening continually occurring you would not be here reading this word. Your experience of the flow of clock time through the present moment is your direct experience of the fundamental process of the universe occurring within your own being.

The flow of clock time is not really so much a separate independent entity that carries individual things along with it but is simply the rate at which processes occur. There is no independent flow of time other than the rate at which processes occur. Thus clock time is not really a thing in itself but is abstracted from the rates at which processes occur.

When processes occur they always happen at fixed rates relative to each other and those rates are what we call the flow of time. Thus the specific rate at which processes occur in any given location is referred to as the local clock time rate at that location.

What this means is that the clocks that measure clock time are simply devices designed to run at standard rates and they measure time by comparing the rates of other processes to their own standard rate based on the principle that local processes always run at fixed rates relative to each other. So clock time is always the rate of some process relative to the rate of some other process. It's intrinsically relative rather than absolute.

For example our personal experience of the flow rate of clock time is relative to the rate of our biological clock, which can vary slightly depending on circumstances. This is why we may experience differences in the rate at which clock time seems to flow at different ages and in different emotional situations.

This also means that without some process occurring there is no flow of time. Thus before the big bang there was presumably nothing happening and thus no flow of clock time and clock time wouldn't have existed (Hawking, 1998).

So we can say that the flow of clock time is simply a term for the rate at which the processes of the universe are being computed at any particular location. If we think of the universe as a computer that continually computes all the processes that are happening within it we could say the processor rate of that computer would determine the clock time rate of the processes it was computing.

Keeping this in mind it's still very useful to speak of the flow of clock time as if it were some actual independent thing even though we are really just talking about the rate at which processes are occurring.

Thus we can say that the present moment is the manifestation of the presence of reality, and the clock time that flows through the present moment is simply the rate at which the processes of reality are happening. These two points explain the fundamental nature of time.

TIME AND RELATIVITY

One of the great discoveries of relativity is that clock time flows at different rates in different circumstances. Specifically in the presence of spatial motion clock time slows. This is referred to as 'time dilation' in relativity and has been extensively verified by observations and must be taken into consideration in space flight and GPS computations (Wikipedia, Global Positioning System).

This means that the rate at which things happen depends on how much relative spatial motion they are experiencing. Things moving faster in space will have their time slowed, and things moving close to the speed of light will have their time slowed to a crawl. However nothing can move faster than light so the speed of light is an absolute limit to all motion in the universe.

The slowing of time with spatial velocity is given by a simple equation called the Lorentz transformation (Wikipedia, Lorentz transformation).

$$\frac{dT}{dt} = \sqrt{1 - \frac{v_x^2}{c^2}}$$

In English this says that the relative time rate of a clock moving with spatial velocity v_x slows proportional to v_x^2 divided by c^2 where c is the speed of light. Thus since most spatial velocities are very small compared to the speed of light, time doesn't slow very much until spatial velocity begins to approach the speed of light. This is why we don't typically see clocks slowing as they move around on Earth. However the slowing is measureable aboard the ISS and spaceflights in general and for any process that has a significant spatial velocity.

THE STc PRINCIPLE

Now there is a largely unrecognized fundamental principle that underlies relativity from which the Lorentz transform is derived [1]. This can be called the STc (space, time, speed of light) Principle and states that the combined space and time velocities of everything in the universe is always equal to the speed of light c.

The equation for the STc Principle is

$$v_x^2 + v_T^2 = c^2$$

which states in English that the square of velocity in space v_x plus the square of the velocity in time v_t is always equal to the speed of light squared.

Note that the combined velocity through space and time is a vector sum rather than the simple addition of their speeds. This is because time is considered a 4th dimension orthogonal to the x, y, and z dimensions of space. Thus by the Pythagorean principle of vector addition it's the square root of the sums of the squares of time and space velocities that's always equal to c [1].

To understand the STc Principle we need to understand what is meant by velocity through time. Relativity expresses velocities through time as *relative* clock rates times the speed of light. Multiplying by the speed of light puts velocity through time in the same units as velocities through space and enables them to be correctly compared as velocities through different dimensions of a single 4-dimensional geometry.

By definition the relative clock rate of an observer's own clock is 1. So multiplying 1 times the speed of light gives c as the velocity through time of an observer and his clock. Thus every observer is always moving through time at the speed of light according to his own clock. The speed of light is actually the velocity in time of everything in the universe including ourselves on its own comoving clock. We, and everything in the universe, are always moving through time at the speed of light on our own clocks. This is what we experience as the passage of clock time.

By definition observers don't move relative to themselves. This is why all the c spacetime velocity of every observer is completely through time on its own clock. Thus c is the speed of time on all observers' own clocks.

Relativity tells us that a clock moving in space relative to us will seem to be running slower than our own clock. Thus two observers moving relative to each other will both see each other's clock running slower than their own. The amount of slowing depends on the relative spatial velocity and is given by the STc Principle equation.

The STc Principle states that the combined vector velocity through space and time of everything in the universe is always equal to the speed of light, c. This is a consequence of relativity that is well known to scientists though usually viewed as a mere curiosity (Greene, 1999, 2005). However it's actually a fundamental principle with profound consequences. It means that every clock in the universe runs slower in time proportional to its velocity through space so that the combined spacetime velocity through time and space of everything in the universe always remains equal to c.

This is one of the fundamental conservation principles of reality. It means that the total velocity of everything in the universe is always c, and this velocity is distributed between velocity in space and velocity in time. Thus if velocity in space increases the velocity in time must decrease so that their vector sum always remains equal to c. Thus c is the fixed velocity of all processes in the universe. Everything in the universe must always move with velocity c no matter how that velocity is distributed between velocity in time and velocity in space. This is a fundamental law of the universe. The STc Principle is the single essential key to really understanding relativity and its effects including how time works.

In a computational universe the STc Principle can be envisioned as the consequence of a fixed number of processor cycles used to compute the happening of every process in the universe. Processor cycles go first into computing the spatial velocity of a process. Then the remaining cycles are used to compute the internal evolution of the process, which manifests as its clock time rate. In this manner the processor cycles that compute each process are distributed among computing velocity in space and velocity in time so that their vector sum is always c. This processor model is explained in detail in my book on Universal Reality (Owen, 2016).

The rate of internal change of state of any process manifests as its associated clock time rate, so the clock time rate slows as a function of the computation of relative spatial motion. This automatically manifests as the STc Principle, which describes how clock time works for all processes. The processor cycles of happening are distributed between computations of relative spatial motion and the internal clock time rates at which processes occur. This is how the STc Principle emerges from the computation of processes in a computational universe.

The STc Principle means the speed of light is actually the intrinsic velocity of spacetime and light just happens to move at that velocity through space because it has no internal processes to be computed and thus no internal velocity through time (. If light had a comoving clock its hands would never move. Thus understanding the STc Principle explains why light itself always has to move through space at the speed of light. It's because it isn't moving at all in clock time so all its c spacetime velocity is through space.

The 3×10^8 meters/sec value of c in our universe provides enough time for things to happen and enough space for things to happen in. If the speed of time was zero nothing could ever happen, and if it was infinite the whole history of the universe would be over before it began, so a viable universe requires a reasonable finite non-zero value of c to be encoded in the complete fine-tuning of the universe.

The value of c must also be quite large, as it is, relative to typical velocities through space. If it weren't the spacetime dilations of mass-energy would produce gravitation so intense as to crush all possible material structures, and routine spacetime distortions so great as to make ordinary processes unintelligible.

The STc Principle not only underlies most of special relativity but also provides a firm physical basis for the apparent mystery of the arrow of time and confirmation of a privileged present moment by relativity itself, a fact that neither Einstein nor most other physicists seem to have recognized. This is explained in those sections below.

The STc Principle applies only to clock time; it doesn't apply to P-time, which has no observable rate since it's an intrinsic aspect of computational reality, namely the processor cycle rate that is fixed throughout the universe.

TIME & GRAVITATION

Relativity tells us that time also slows in gravitational fields which suggests gravitation might be some form of spatial velocity. In fact it's easy to model gravitation as a field of vibrations in space whose intrinsic velocity is equivalent to their gravitational force (Owen, 2016).

This is a simple and elegant theory that allows both linear velocity and gravitation to be seen as two aspects of the same phenomenon. Basically the theory says that all forms of mass and energy are different forms of spatial velocity. This has the added advantage of explaining how all forms of mass and energy can be conserved as science has shown they are. It's easy to understand how various forms of the same thing can be converted from one to the other and in fact the only way anything can be conserved is if it's all forms of the same thing. This is all explained in detail in Universal Reality (Owen, 2016). Thus we discover the underlying reason for the conservation of all forms of mass and energy. They are all forms of spatial velocity.

Thus the conservation of mass and energy is simply the conversion of equivalent amounts of spatial velocity among its various possible forms. The linear velocity of motion, the wave frequency velocity of electromagnetic radiation, and the vibrational velocity of gravitating mass are all interconvertible forms of spatial velocity. This insight makes understanding time and its relativistic effects much easier to understand.

To understand the STc Principle in the context of gravitation we simply model a gravitational field as a field of intrinsic velocity in space. Thus an object seemingly at rest in a gravitational field actually experiences an intrinsic spatial velocity because it's in a velocity density field.

A gravitational field is a field of intrinsic velocity with strength inversely proportional to the square of the distance from the mass that produces it. Thus a gravitational field has an intrinsic velocity gradient extending out from the mass and every point in the field has a resulting velocity vector towards the mass because the intrinsic velocity of the field is greater closer to the mass than away from it and the difference at any point produces a velocity vector pointing towards the mass.

Thus an object at rest in the field experiences a velocity vector towards the mass that its inertial motion tends to follow. This is the source of what we call gravitational attraction. There is no actual attraction in the sense of a pull. It's just a matter of an object in inertial motion traveling through space along its velocity vectors.

We confirm a gravitational field as a velocity field in our direct experience as the velocity we experience falling towards a gravitating mass along the velocity vectors it produces. And the force of gravity we experience standing on the surface of the earth is actually an acceleration against this inertial motion due to the surface of the earth blocking our motion. This is in agreement with Einstein's Equivalence Principle that states that what we feel as a gravitational force is actually an acceleration (Wikipedia, Equivalence principle).

So to understand how the STc Principle works in a gravitational field we simply add the intrinsic velocity density of the field to any linear velocity an object has. The velocity through time of the object then slows due to the combined spatial velocity of linear velocity and the intrinsic gravitational velocity of the field. This explains why both linear velocity and gravitational fields produce time dilation. They are both forms of spatial velocity that slow the velocity of time by the STc Principle.

Thus a clock at rest in a gravitational field will observe a clock at rest in empty space running faster because its own clock rate is slowed by the intrinsic velocity of the field. The clock in empty space will likewise see the clock in the gravitational field running slower for the same reason. Both observers agree on this effect; thus it's an actual relativistic effect and the difference in elapsed time produced is permanent.

The velocity density model explains the relativistic slowing of time by gravitational fields in the same way time is slowed by linear velocity. Gravitational time dilation and the time dilation of linear motion are both due to increased spatial velocity slowing velocity in time in accordance with the STc Principle.

Thus the total spatial velocity of an object is its linear velocity plus the intrinsic velocity of any fields it's in. This gives its total spatial velocity, which combined with its temporal velocity must always equal c.

Thus we get a very simple way of understanding general relativity that works for all observers in all situations. An observer in a gravitational field just needs to add the intrinsic velocity of the field to any linear velocity he may have to correctly calculate his time dilation compared to an observer at rest in empty space.

This velocity density model of gravitation is equivalent to the curved spacetime model of general relativity and predicts exactly the same effects but is actually superior because it provides an actual mechanism for those effects that is lacking in relativity. Relativity provides no mechanism for how the presence of matter curves spacetime, it just does. However in the velocity density model mass and its gravitational field *actually are* fine spatial vibrations whose spatial velocity works seamlessly with the STc Principle. Gravitational mass and linear velocity now become two aspects of the same thing as the conservation of mass and energy suggests they should have been all along.

And not only that the velocity density model has the additional advantage of modeling space as we actually see it, as a flat uncurved Euclidean space rather than the curved space of general relativity which we never observe. In the velocity density model space is flat and uncurved as we actually see it, it just contains areas of intrinsic velocity density around masses because masses actually are fields of velocity density in space.

THE VELOCITY FABRIC OF SPACETIME

The STc Principle is a universal fundamental principle that tells us the speed of light is actually the fundamental velocity of spacetime itself, and space itself can be considered as nothing more than a field of intrinsic velocity whose value is c at every point. In this view the fabric of space itself consists entirely of intrinsic spacetime velocity.

The amount of that intrinsic velocity that consists of intrinsic *spatial* velocity is the strength of gravitation at any point in space. Correspondingly the amount of total velocity at any point that is not the intrinsic spatial velocity of a gravitational field is the intrinsic velocity of time at that point. This gives us a very neat and elegant picture of the basic unity of mass-energy and spacetime that automatically obeys the rules of general relativity which can be called the METc (mass, energy, time, speed of light) Principle (Owen, 2016).

Thus the speed of light should actually be understood as the intrinsic velocity of spacetime, or more fundamentally as the intrinsic velocity of mass-energy plus time at any point of computational space. In this view space itself is a field of mass-energy, a field of intrinsic spatial velocity with its density corresponding to the strength of gravitational fields present.

In this view the masses and other charges of particles are simply little units of spacetime that have broken away from the background and crystallized around their particle components. Thus they become free to move on their own relative to the background space. All particle masses and other forms of energy are actually forms of concentrated space velocity free to move against the background field of intrinsic spatial velocity. This theory is explained in detail in Universal Reality and the concept of particles as excitations of space is also consistent with quantum field theory (Wikipedia, Quantum field theory).

If a unit volume of space contains strong mass vibrations the distance across it is further because the ups and downs of the vibrational waves must be traversed. This is completely equivalent to the curved spacetime model of general relativity because its curves could be compressed into vibrational waveforms in a flat Euclidean space and if they were stretched out again the resulting space would be curved.

In this model a mass is a field of minute vibrations in a flat Cartesian space centered on the massive particle(s) producing it. This is much easier to visualize than the curved spacetime of general relativity. The beauty of the velocity density vibrations model is its flat Cartesian space is very easy to understand and work with and presumably much simpler for the universe to compute.

In the vibrational space model the speed of light is the same for all observers everywhere in the universe. Light appears to move slower across vibrational space because actual distances across it including the ups and downs of the vibrations are much greater than they appear. Thus the actual distance traversed by light up and down the vibrations is such that light always does travel at c everywhere in the universe even in gravitational fields.

In the standard interpretation of general relativity this is due to the curvature of space in gravitational fields. Thus light beams are actually always traveling at c but when they have to traverse the curvature of space the distance traveled is greater than it appears so light appears to move slower than c to a remote observer. We don't directly see the curvature of space produced by gravitational fields because light travels along it but the actual distance is greater than the nominal distance between two points in a gravitational field. Vibrational density spaces model gravitational fields as flat Cartesian spaces as we actually see them rather than the curved spaces general relativity uses and this is another significant advantage of Universal Reality.

Thus we actually see the sun very slightly larger than we would if it didn't curve space near it and it's actually slightly farther away than its nominal distance measured along the curved space of its gravitational field. This effect is proportional to gravitational strength, which is why light appears to slow to zero speed as it nears a black hole and images of things pile up at the event horizon and fade out as they cross it (Wise, ?).

The STc Principle and the equivalence of gravitational mass-energy and spatial velocity, the METc Principle, are the two keys to understanding general relativity. And our understanding is greatly improved by replacing the spacetime curvature model with an equivalent intrinsic velocity density vibrational space model.

Even though time travels at different rates depending on different relativistic conditions, there is a hypothetical standard clock time rate for the universe, which is the maximum rate clock time can flow. This is the time rate of stationary clocks in deep space far from any gravitational field though there is nowhere this is strictly true.

This is the clock time rate of empty space, which is likely related to the intrinsic velocity of the zero-point energy of the quantum vacuum (Wikipedia, Zero-point energy). Since all forms of mass-energy produce gravitational effects the zero-point energy also produces a gravitational effect. And since all gravitational fields are equivalent to fields of intrinsic velocity, and the amount of intrinsic velocity sets the balance of space and time velocities there should be a relationship between the values of c, G (the gravitational constant), the zero-point energy, and perhaps the total mass-energy of the universe.

Thus the clock time rate of time in empty zero-point energy space is the standard baseline clock time rate of the universe. It would be the maximum possible clock time rate and all other clock time rates would be slower proportional to the presence of additional spatial velocity. Nevertheless all observers measure their own proper time rates as the speed of light on their own comoving clocks even if they are in a gravitational potential or moving through space as explained above.

ACTUAL & OBSERVATIONAL TIME DILATION

One of the important aspects of relativistic time is the difference between what can be called actual and observational time dilation. Relative spatial motion always produces time dilation but whether that time dilation is actual or just observational depends on what it's actually relative to. If the motion is relative to an actual world line through the background fabric of spacetime the time dilation is actual. Actual means that all observers agree on the amount of time dilation and the difference in elapsed time produced is permanent.

On the other hand two observers moving relative to each other will both observe the other's clock running slower by the same amount. The time dilation is reciprocal and vanishes with no permanent effect as soon as the relative motion stops. In this case the time dilation is merely observational.

In all cases of relative motion there is always some relative motion with respect to the spacetime background by one or both of the observers. And only the amount of relative motion with respect to the actual spacetime background produces actual time dilation. All the rest of the relative motion produces only observational time dilation.

Relativity itself has a hard time explaining the difference between actual and observational relativistic effects, as the founding principle of relativity is that all frames are relative and none are privileged or absolute. This is a major blind spot of relativity that is easily solved by a computational approach to reality.

The only way the difference can be explained is if there is an absolute spacetime background roughly aligned with the aggregate mass of the universe with respect to which the spatial velocity that produces actual time dilation is relative. In a computational universe this is simply the computational space in which the dimensionality of the universe is actually computed.

This approach has the added benefit of explaining what actual rotation is relative to and solving the problem of Newton's bucket; what actual rotation is relative to. This is explained in detail in Universal Reality (Owen, 2016).

Thus the takeaway is that for the observed effects of relativity including time dilation to actually make sense there has to be an absolute background space with respect to which the spatial motion that produces actual relativistic effects is relative. Combined with present moment time this gives us a universal background spacetime with respect to which all actual relativistic effects occur.

The equations of relativity are the same for both actual and observational effects. The only difference is that actual relativistic effects produce permanent effects that all observers agree upon, while observational relativistic effects are seen differently by different observers and vanish when the relative motion ceases.

TWO KINDS OF TIME

We all directly experience the fact that there is a present moment through which time flows, and we know from relativity that time can flow at different rates through the present moment. The only way this most fundamental experience of our existence makes sense is if there are actually two different kinds of time. As strange as it may originally seem there is simply no way around this obvious fact.

There must be two different kinds of time, the time of the present moment and clock time which runs at different relativistic rates within a single universal present moment. The time of the present moment is universal and absolute and common to everything in the universe. And there is clock time, which flows through the universal present moment at different rates depending on the local presence of spatial velocity. Clock time flows through the current present moment at different rates but the universal present moment is common to all observers throughout the entire universe.

The fact there are two kinds of time is conclusively demonstrated by the established fact that relativistic observers always reunite in the exact same present moment even if their clocks read different clock times.

There is no doubt at all that there are two kinds of time and it's totally amazing that no one had recognized this obvious truth until I first pointed it out (Owen, 2007, 2009). This serves as an excellent example of the blindness of science to obvious facts that somehow just don't register in current memes even when perfectly consistent with established science.

Obviously this proposal is controversial and requires good evidence to be taken seriously. However it's pretty straightforward to demonstrate. First, to prove clock time and the present moment are two separate kinds of time we need only demonstrate there is a single universal present moment within which clock times vary since it's already an experimentally proven and widely used fact that clock times do run at different rates according to relativistic conditions. Thus if we can demonstrate this occurs in a common universal present moment, then clock time and the present moment must indeed be two different kinds of time.

The existence of a common present moment throughout the universe is not a new or strange idea. It was the standard accepted view of time throughout history until the advent of relativity. Clock time was thought to be flowing at the same rate throughout the universe and the present moment was thought to be the common universal present moment of *clock time* rather than a separate kind of time. The present moment was the current reading of a universal clock, which ran at the same rate throughout the universe. Thus there was only a single kind of time, clock time, and the present moment was the present moment of this universal clock time.

But with the advent of relativity it became clear that clock time didn't flow at the same rate everywhere so there couldn't be a universal *clock time* that was the same everywhere. So because time was still considered a single entity the newly variable clock time was still thought to be the only kind of time and the very notion of a universal present moment inconsistent with this view was wrongly discarded. The notion of two separate kinds of time to reconcile clock time with the present moment doesn't seem to have occurred to anyone until now.

Even though the present moment is a central experience of our existence scientists post relativity couldn't come up with any explanation of what the present moment was, and it was either ignored or even denied and replaced with truly outlandish theories such as block time in which all times exist simultaneously and there is no special present moment [2].

The main reason for the current ignorance of the present moment in modern physics is most likely that it has no obvious measure and physicists have the unfortunate habit of ignoring or denying the existence of anything without measure even though neither consciousness, nor existence, nor the present moment have measure and they are our three most important and fundamental experiences of reality.

So the important insight of this new understanding is to retain a universal present moment and recognize that relativistic clock times run at different rates within it. This is a very simple and reasonable insight in accord with our direct experience of reality, totally consistent with relativity, and it has profound consequences in our understanding of reality. And it also turns out this concept of two kinds of time is even an implicit though totally unrecognized principle of relativity itself without which relativity doesn't even make sense.

Every comparison of different clock times in relativity only makes sense if there is a common present moment in which the comparison takes place. There must be a common present moment that serves as a common background reference. Thus a common universal present moment in which *relativistic* comparisons can be made and shared is a hidden and completely unrecognized assumption of relativity used by physicists all the time but which many actively deny! This is one of the great blind spots of modern science.

For example if two space travelers with different clock times were really in each other's pasts and futures they would be completely unable to compare their clocks. They can compare the different readings of their clocks only because they and their clocks are both in the same present moment. Their present moment is the same but their clock times are different, thus it's clear there must be two different kinds of time.

When one of two twins embarks on a relativistic space journey they part in a common present moment. They then each continuously experience their separate existences in a present moment throughout the duration of their separation. And when they meet again they always meet in a present moment common to both even though their clocks now read different times. So there is no reason whatsoever to think the present moment they both experienced during the entire duration of the separation was not the same present moment for both. Their ages and clock times are now much different but their present moments never lapsed and must have always been the same even though their clocks were running at different rates.

Thus it's reasonable to assume, with no evidence at all to the contrary, that there is a single common universal present moment throughout the entire universe, and to assume that every observer in the universe is always in the same current universal present moment as everything else. Thus all that exists, the entire universe, exists in the single common current universal present moment.

This is really quite obvious, but it has been by far the most contentious aspect of the theory of Universal Reality with various arguments being raised against it. Some of these arguments have been based on a specific misunderstanding of the theory, that it violates the relativity of simultaneity in which different observers can have different valid observations of whether two events occur at the same *clock time* or not (Wikipedia, Relativity of simultaneity).

But Universal Reality accepts and incorporates all the equations of relativity including those of the relativity of simultaneity. The relativity of simultaneity correctly describes the behavior of *clock time*, but says nothing about *present moment time*. Non-simultaneous clock times are always compared in the same current present moment so this argument has no bearing on whether there are two kinds of time and can be disregarded.

For some very strange reason the existence of two separate kinds of time is beyond the comprehension of many otherwise intelligent people. Even though it's really quite a simple, logical and straightforward idea many people have great difficulty wrapping their heads around it.

SOME THOUGHT EXPERIMENTS

There are a number of useful thought experiments that more

clearly demonstrate a universal present moment. To prove our point we must demonstrate both that there is a common universal present moment for all observers stationary with respect to each other and in motion or acceleration with respect to each other.

We already know that whenever any two observers are spatially collocated they both experience the same common present moment because they can communicate more or less instantaneously to confirm this. And this is true whether they are moving or stationary with respect to each other.

Consider first a universe completely filled with stationary observers packed tightly together like sardines. We already know every one of them is continuously in the same present moment as the adjacent observers on all sides and this is true for all observers in the universe even if all their clocks are running at different rates in different gravitational potentials. Therefore every observer across the entire universe must be in the same common present moment and this present moment must be universal.

This is a simple proof that there must be a single common present moment throughout the universe that holds whether the observers are stationary or in relative motion because it also holds if the observers are moving relative to each other and just happen to transiently assume the packed sardine configuration when the experiment is performed.

This proof also holds whether or not there are gravitational fields involved. The clocks of observers in gravitational fields will be running slower and across the entire universe all clocks will be running at varying rates in the same universal present moment and this will be agreed by all observers as each confirms their existence in the same current present moment with all those adjacent.

Now the counter argument might be raised that there is some slight difference in present moments across the entire universe too small to be noticed by spatially adjacent observers. But what could a difference in present moment even mean? The present moment most certainly doesn't correspond to differences in clock times because we already know different clock time rates exist in the same present moment. So what would such a difference in present moments amount to? What would cause it and why would that difference be magically erased when separated space travelers meet with different clock times? If their present moment times were different during their separation how could they become the same again when they met? What would be the mysterious mechanism involved? It wouldn't make sense.

Now consider another thought experiment involving two space travelers who part with synchronized clocks, accelerate at different velocities in different directions and periodically redirect to cross paths. We know they are in the same present moment whenever they cross paths so take this example to the extreme and assume they accelerate with enormous and varying velocities but turn and cross paths more and more frequently until the interval between meetings approaches zero. Now they are crossing paths every minute fraction of a second. And every time they cross paths they both confirm they share the same present moment, and no matter how they change their accelerations and how much their clock times vary this is always true.

So it seems outlandish to assume that somehow in the minute fractions of a second they are separated their present moments somehow become different. Their clocks can run at very different rates and read differently every time they cross paths but this is always happening in the same shared present moment. And in fact they both can confirm to each other their clocks are running at different rates every time they cross paths in the same present moment.

This is also confirmed by ground communications with the International Space Station. Time aboard the ISS progresses at a measurably slower rate than on earth because of the velocity it's traveling along its world line (Wikipedia, Time dilation). Yet the ISS is in continual contact with the earth at all times. The ISS and earth are both in the same shared present moment at all times even as their clock times run at different rates. True there is a slight communication delay but this delay occurs in the common present moment.

Now imagine the ISS accelerating enormously but maintaining the same circular path around the earth. Clock time and all physical processes aboard the ISS will begin running perceptibly slower from the point of view of earth observers and vice versa, and this can be confirmed by continual mutual communication. The slight time delay in communications remains the same confirming it's irrelevant. Observers on both earth and the ISS would now see each other's clocks and people moving in slow motion but they would both continually communicate this fact back and forth in the same shared present moment even though each would hear the other speaking slower.

Their times would certainly seem very strange to each other but since they remain relatively close throughout they both would be able to continually observe this mutual strangeness in the same present moment. They would both continually observe each other's clock times running at wildly different rates within the same shared present moment.

So it's clearly possible for two observers to actually watch each other's clocks run at different rates in the same present moment in some cases. This can only be true if the present moment and clock time are two different kinds of time. Science must be based on observation and the present moment and two different kinds of time are clearly observable facts.

In light of these thought experiments it seems undeniable that there is a single present moment common to the entire universe that is completely different than clock time, and therefore there are two completely different kinds of time.

THE PRESENT MOMENT

It's important to clearly understand what is meant by the *same* present moment. There isn't a single present moment that stays the same over all time as clock time flows through it at different rates. Instead there's a separate present moment kind of time that also progresses but at the same rate throughout the universe. The current present moment right now isn't the same as the previous present moment. For things to happen there must be a current present moment that is not the same as the previous current present moment.

The present moment time is the successive present moments of the active process of happening in the observable universe, which continually manifests as a universal present moment in which everything exists, and in which the state of the universe is continually being recomputed. This theory of two kinds of time is completely consistent with science including relativity and with personal experience and informed common sense.

Happening can be considered as a processor that continually computes the current data state of the entire universe. Since all the data of the universe exists simultaneously within the processor of happening the *presence* of existence manifests as a universal current present moment in which everything happens. This current present moment is universal and common to all observers no matter how fast their clocks are running.

Thus happening is the source of the fixed flow rate of present moment time throughout the entire universe. The time of the present moment flows at the same universal rate throughout the universe, but within that universal flow the different relativistic flow rates of clock time are computed locally on the basis of the STc Principle.

However it's not easy to pin present moment time down because it has no intrinsic metric since all clock time metrics are computed within it. Thus present moment time is prior to the computation of any observable dimensionality of time. Happening computes all dimensionality including clock time metrics so it has no observable measure of its own.

If present moment time and computational space are the source of all dimensional space and time metrics they can have no observable metrics of their own. Thus they are pre-dimensional in the same sense that a numeric representation of dimensional data in a computer program is not itself a dimensional structure. Thus a computational universe is intrinsically non-local and events across the entire universe can be computed simultaneously irrespective of the finite speed of light. This has important relevance to the apparent problem of quantum non-locality (Owen, 2016).

Thus there is a present moment time that progresses with happening but no intrinsic associated metric other than the various clock time rates that it locally computes. So two distinct kinds of time do exist and both progress, but only clock time has a measurable metric associated with it. Thus present moment time can only be measured in terms of the clock times it produces.

There is also another way to get a sense of the dimensionality of present moment time in terms of its effect on cosmological geometry as explained in the section on the Hyperspherical Universe. In fact the concept of a separate present moment time is doubly important because it immediately nails down the previously uncertain geometry of the observable universe.

P-TIME

If we call the time of the present moment P-time, then every actually occurring event takes place simultaneously for all observers at the same P-time throughout the entire universe. In other words if an event takes place in the present moment for one observer, it also takes place at the exact same P-time for all observers. This will always be true no matter how different the clock times and clock rates of various relativistic observers are. This universal simultaneity of P-time is what allows relativistic observers to compare their different clock times in the same present moment.

This P-time simultaneity has nothing to do with the relativity of simultaneity of *clock time*. Different observers can still have different views of the clock time simultaneity of events due to the finite speed of light between observers and events. However the entire observable universe, including its different clock times, is continually recomputed in the same universal P-time moment.

There is another good argument for a universal P-time, but we first need a couple of basic definitions from relativity. Relativity defines *proper time* as the clock time reading of an observer's own comoving clock, the clock on his wrist or wall. *Coordinate time* is the time an observer sees on another observer's clock and is a measure of the intrinsic rate of processes associated with that other observer from the perspective of the first observer. An observer's measurement of another observer's coordinate time is always by comparison to his own proper clock time, and that measurement, like all observations, is always made in the first observer's present moment.

The proper times of all clocks are continuous; there are never any gaps in the flow of their P-time. There is never any moment that proper time is discontinuous during the separation of observers or at any time for that matter. Thus there must always be a one-to-one correspondence of proper clock times for any two observers. There must always be some proper time reading on one observer's clock for every proper time reading on the other's clock whether they are different or not.

This gives us an additional means of testing the universal present moment theory. Every observer can specify his progressive P-times in terms of his own proper clock time readings. There is always a proper time clock reading for every current present moment since both clock time and P-time are continuous. For example, observer A can say he was tying his shoes in his present moment when his proper time clock read 12:00 AM. Every observer's proper time clock reading serves to uniquely identify his own current present moment P-time at that time, and this is true of all observers. For every current present moment of every process in the universe there was always a corresponding proper time reading that can be used to denote it.

Therefore if there is a single universal P-time that flows at the same rate for all observers so they all remain in the same common universal present moment, then there must also be a unique one-to-one relationship between the proper times of all observers even when they differ. There must be one and only one proper time on every observer's own clock that corresponds to every single P-time they all shared when their clocks read those proper times. If there isn't then the universal present moment theory is falsified, but if there is it's confirmed.

Simply stated, for every present moment proper time of any one observer every other observer must have been doing something at that exact same present moment at their own corresponding proper times, no matter how differently their own proper time rates may have been flowing relative to one other.

In many cases this proper time correspondence can be calculated and confirmed. If the space traveling twins exchange flight plans before they separate the one-to-one correspondence of their proper times throughout their separation can be calculated. Each will know the complete relativistic history of the other and thus know exactly how much proper time has elapsed on the other's clock for any proper time on his own clock. Each will know what the other's clock is reading at every moment on his own clock. To be absolutely clear this *calculation* of the current *proper time* of the other twin is not the *coordinate time* that would be *observed* on the other twin's clock. It's the *proper time* of the other twin's clock, which is not generally observable but which can be calculated.

Thus it's always possible for any observer who has knowledge of the relativistic circumstances of any other observer to calculate what proper time reading of that other observer correlates to each of his own proper time readings. There is always a one-to-one correspondence of proper time readings between any two observers that tells them what proper time of one corresponded to the proper time of the other in their common present moment even when they are separated in different relativistic circumstances with different clock time rates.

However in the general case of any two observers in the universe, where an initial proper time correspondence can't be determined or the relativistic history of the other observer is not known, it may be impossible to calculate the current proper time correspondence even though it's certain one must exist. However if any observer in the universe can determine the relativistic variables of any other observer as well as his own he can calculate the proper time rate of that observer relative to his own proper time rate and confirm the existence of a common shared P-time.

Thus it's easy to show there will always be a one-to-one proper time correspondence between any two observers in the universe, and this is all that's necessary to conclusively demonstrate a universal P-time present moment common to all observers. It's sufficient to note that time is continuous for all observers thus every observer in the universe must be doing something at every proper time moment of any other observer's clock. Whatever is being done will be done in the exact same common universal present moment of all existence in the universe because the present moment is the only time that anything can occur because it's the only moment that exists and the only locus of reality.

The clock times of different observers can flow at different rates through this common present moment, but there is always a one-to-one correspondence between the proper times of any two observers throughout the entire universe. This is consistent with Universal Reality's proposal that the common universal present moment of existence is all that exists, and is the current moment of happening in which all the computations of the universe take place. It is the current universal P-time tick of the entire computational universe.

Therefore the proper times of observers can be used to notate the passage of P-time, and their correspondences, when they can be determined, can be used to establish identical P-times among observers even though P-time has no intrinsic metric of its own. And if all else fails any two observers can simply communicate their current relativistic conditions and proper times to enable their P-time simultaneity to be calculated.

THE ARROW OF TIME

One of the perennial mysteries of science has been the source of the arrow of time; the fact that time continuously flows in the forward (by convention) direction. Many scientists have sought the source of the arrow in vain, often mistakenly attributing it to entropy (see the section on 'Entropy and Time' for why this isn't correct), but the explanation is quite straightforward and a simple consequence of the STc Principle.

The STc Principle states that everything continuously travels at the speed of light through time on its own clock. Therefore it's quite clear that time must be experienced as flowing in a single direction by every observer and that every observer's own clock will always appear to have its hands moving at the speed of light in the same direction as all others.

Thus the STc Principle itself is the actual source of the arrow of time, and puts the arrow of time on a firm scientific basis. The arrow of time is a direct implication of the theory of relativity; another completely unrecognized fact of modern physics, and even actively denied by scientists who don't understand it.

CONFIRMING A PRESENT MOMENT

Amazingly the necessity of a privileged present moment of time distinct from all other moments of time is another direct consequence of the STc Principle, again completely unrecognized by most physicists. In fact many physicists continue to deny the existence of a present moment mistakenly believing it's inconsistent with relativity. It is not. It is actually *required* by relativity and a direct consequence of the STc Principle that underlies special relativity.

By the STc Principle everything continuously travels through spacetime at the speed of light. This means everything must always be at one and only one point in time, and that point is obviously the current present moment of its actual existence. Thus the STc Principle requires the existence of a privileged present moment for all observers that is the current time of their existence, and the *only* time they are actually at, the time that defines their now.

Thus relativity itself absolutely requires a present moment that progresses in clock time and conclusively falsifies the nonsensical 'block universe' hypothesis in which all times exist at 'once' as a single static structure (Price, 1996, 12-13, 14, 15-16), (Wikipedia, Eternalism (philosophy of time)).

This present moment required by relativity is the same present moment we have already identified as the presence of existence that intrinsically manifests as a single universal present moment in which everything exists. Thus Universal Reality is clearly consistent with and confirms a proper understanding of relativity in this respect.

Because existence exists it must have a presence and its presence manifests as a universal present moment in which everything exists. The universal present moment is simply the presence of existence and encompasses all that exists and there is no before or after it that actually exists. The presence of existence is the source of the present moment that is now confirmed by the STc Principle to be consistent with relativity.

Physicists who deny the existence of the present moment should remember it's the most fundamental and persistent of all observations, and that the role of science is to explain observations, never to deny them.

THE HYPERSPHERICAL UNIVERSE

Relativity has revealed that we live in a 4-dimensional universe consisting of 3 dimensions of space and 1 dimension of clock time and this is clearly correct. However relativity itself hasn't been able to discover the actual geometry of our universe because it hasn't recognized the present moment as a second kind of time. The unfortunate result is science has no clear picture of the overall geometry of the universe and tends to fall back on clearly inaccurate expanding tube like images with flat surfaces and edges (Wikipedia, Metric expansion of space, #Topology of expanding space).

Modern science is positively schizophrenic when it comes to the notion of a universal time. First physicists adamantly deny the concept of a present moment and the notion of a common universal time, and in the next breath they tell us the universe is 13.8 billion years old and that's true for every observer in the universe. Then they engage in all sorts of genuflections to try to reconcile these clearly contradictory views.

The obvious cosmological geometry of the universe is a 4-dimensional hypersphere where the 3 surface dimensions are our 3 dimensions of space and the radial dimension is the time dimension from the surface back to the center point of the big bang.

However this doesn't work if the radial dimension is clock time because clock time can run at different rates within the surface and there could be no single consistent radial time dimension. This is why physicists haven't been able to discover the true hyperspherical geometry of the universe.

However if we recognize the separate time of the present moment and take the radial dimension as P-time instead of clock time the hypersphere works quite well because the P-time rate is fixed across all regions of the surface and we get a single consistent P-time radial dimension for the universe. The obvious 4-dimensional hyperspherical geometry now makes perfect sense.

Everything that exists, the entire universe, exists in the current present moment surface. The past no longer exists and all the past onion-like layers of the hypersphere where the surface used to exist correspond to the history of the universe over past P-time back to its origin at the point of the big bang at the center.

Thus at each P-time tick the entire surface of the actually existent universe is recomputed and a new very slightly larger surface in the next present moment is created. Thus the 3-dimensional spatial surface of the universe continually expands as P-time progresses.

As the progression of P-time continually expands the surface local clock time rates are computed across the surface depending on the presence of spatial velocity according to the STc Principle. The 3-dimensional surface is the current present moment fabric of space and consists of a spacetime velocity at every point equal to the speed of light c. Any intrinsic velocity of a gravitational field at any point automatically reduces the velocity of time at that point so their vector sum is always c.

Thus the universe takes the form of a closed finite 3-dimensional hyperspherical surface in the current present moment with positive curvature and no edges at the largest scale. It cannot be infinite because nothing actual can be infinite because infinity is not an actual state or fixed number but a never-ending *process* of continual addition. Infinity is a useful mathematical concept but nothing actual can be infinite.

Nor is there any reason to believe the universe is not a closed continuous surface and has edges. How could the point universe of the big bang develop edges as it inflated? That would tear it apart and it's clearly nonsensical. Traveling in a straight line in any direction across the universe one would theoretically eventually end up at approximately the same place ignoring any local curvatures of space just as one does by circumnavigating the earth ignoring the mountains and valleys.

If P-time is the radial time dimension of a hypersphere then the circumference of the spatial surface of the universe will be a function of its P-time radius, and measurements of the curvature of space should provide a measure of its radius and the P-time age of the universe. Current measurements suggest that 3-dimensional space is fairly flat within its observable volume but a hypersphere is not ruled out (Wikipedia, Flatness problem). A hypersphere also makes sense from the perspective of general relativity, as the mass-energy content of the universe should curve it in on itself at the largest scales.

This hyperspherical geometry should be subject to experimental confirmation since the curvature of space is measurable (Wikipedia, Shape of the universe). It should turn out to have a very small positive curvature. But even if it doesn't that raises doubts but doesn't necessarily falsify the hyperspherical geometry since if the hypersphere is not perfect it could be closed and finite and still contain some areas with greater or lesser or even possibly negative curvature.

Only a small volume of the cosmic hypersphere is visible from any location within it since its spatial surface is uniformly expanding at a rate that exceeds the speed of light beyond a distance called the particle horizon. Because space itself is expanding away from us faster than the speed of light beyond the particle horizon, light can never reach us from there and that area of the universe isn't visible to us. Likewise we are not visible from points beyond our particle horizon because we are beyond the particle horizons of those points.

However the entire current P-time surface of the hypersphere including all regions beyond the particle horizon is the whole actual universe since the entire surface is the current present moment of P-time in which the entire universe is recomputed and exists. The entire surface of the hypersphere, the entire universe, is in the same P-time present moment all around its surface, irrespective of particle horizons, and irrespective of the various local rates of clock time.

Note that the particle horizon is an observational as opposed to an actual relativistic effect. Beyond the particle horizon nothing is actually moving faster than the speed of light with respect to the background fabric of spacetime. Processes evolve normally just as they do in our area of the universe. The absolute dimensional background of computational space in which all processes are computed extends around the whole surface of the hypersphere with no interruptions or anomalies. It's only when processes near some observer's particle horizon are observed from afar that anomalies appear to exist. But all such anomalies are anomalies of clock time observations and have nothing to do with P-time, which continues to recompute the entire surface with each universal P-time tick.

INFLATION & THE HUBBLE EXPANSION

The hypersphere is a neat and elegant model of the universe but it initially seems to have two problems. First the rate of the P-time radius expanding the surface is presumably constant but the expansion of the surface seems to have varied greatly over the history of the universe and the current Hubble expansion of the surface appears to be accelerating (Wikipedia, Accelerating expansion of the universe). If the rate of increase of the radius is constant the expansion of the surface should also be constant.

And second if the radius of the hypersphere was only the 13.8 billion years of its clock time age back to the big bang the size of the universe would have to be much smaller and its curvature would have to be very much greater than current measurements suggest.

The solution to these apparent problems is straightforward. Science measures the expansion of the surface in *clock time* but the extension of the radial dimension of the hypersphere is in *P-time* and their rates are not proportional because P-time computes various clock time rates depending on the presence of spatial velocity. For example by the STc Principle if there were a time when the spatial velocity in the universe was much greater than it is now the overall clock time rate of the universe would have been much slower.

In fact there was such a time called inflation. The inflationary period of the universe was an enormous exponential expansion of the universe in the first slight fraction of a second after the big bang (Wikipedia, Inflation (cosmology)). There is considerable evidence for inflation and the theory is widely accepted.

Inflation was an enormous explosion of the spatial velocity of the entire universe. Thus by the STc Principle there must also have been an equally enormous slowing of clock time throughout the entire universe. This means that the overall clock time rate of the universe slowed to almost nothing and clock time barely passed at all during inflation because the same rate of P-time happening was busy computing the spatial expansion instead.

Thus from our current look back rate of much faster clock time, inflation seems to have occurred almost instantaneously in clock time while presumably P-time would have been running at its standard rate. Thus enormously more P-time would have passed than clock time, which means the P-time age of the universe is enormously greater than its apparent clock time age of 13.8 billion years.

This means that the actual radius of the hypersphere is much greater than 13.8 billion light years and its circumference will also be much greater. This immediately allows the curvature of the hypersphere to be consistent with current measurements of the curvature of the universe and the apparent problems with the hyperspherical model can be resolved.

So clearly the changing rates of expansion of the universe in clock time over its history can be decoupled from the presumably uniform rate of extension of its P-time radius. Thus the changes in the rate of Hubble expansion of the universe over its history including its current apparently accelerating rate are consistent with the hypersphere model.

After the initial period of exponential inflation, the expansion of the 3-dimensional space of the universe seems to have quickly decelerated to a much slower rate before beginning to gradually accelerate again. The expansion appears to be still accelerating. This expansion of the universe is called the Hubble expansion after its discoverer, Edwin Hubble (Wikipedia, Hubble's law).

Even though the expansion is imperceptible at local scales, over intergalactic distances it adds up to produce particle horizons equidistant from every point beyond which the expansion exceeds the speed of light and nothing is visible.

So the Hubble expansion is an extraordinarily slow expansion of the fabric of space. At any given point of space it's completely imperceptible and only becomes apparent over intergalactic distances. So any minute changes in expansion rate in the spacetime fabric over time are very easy to reconcile with the presumably constant rate of the P-time extension of the radius of the hypersphere by very slight changes in the overall spatial velocity content of the universe very imperceptibly reducing the overall clock time rate.

Any expansion in the fabric of space is spatial motion and thus slows the local clock time rate. The Hubble expansion is a fairly uniform expansion of the entire surface of the hypersphere and thus the clock time rate of the entire universe is slowed. However that slowing would be locally imperceptible since it occurs point by point across the universe where the expansion is likewise far below the level of possible measurement. The expansion only shows up at intergalactic scales as red shifts as does the slowing of clock time in the greater time it takes light to cross intergalactic distances across expanding space and the stretching of its frequency which is what produces red shifts.

So the continual P-time extension of the radius of the hypersphere does produce the Hubble expansion, but when it's measured in clock time the Hubble expansion can vary because the expansion automatically produces a concurrent slowing of clock time.

It should also be pointed out that though the Hubble expansion appears to be accelerating it's not clear how accurately we even know the expansion rate over time. Due to the finite speed of light we are only able to observe the universe as surfaces of fixed distance and time. So we have no direct observations at all the other distances for any given time, and al all the other times for any given distance. This gives us only a minute sample set of the rates of expansion over the observable history of the universe.

In particular we have no idea at all of what the *current* expansion rate of the universe really is since its totally unobservable, and will remain so until the light from standard candles begins to arrive which will be far into the future.

Standard candles are objects such as Type 1a supernovae, which all have approximately the same intrinsic brightness and thus allow very accurate estimates of distance. Their apparent brightness gives an accurate estimate of distance, and their red shift gives a measure of the recession velocity of space at that distance. However another problem is that type 1a supernovae are quite rare and transient and widely dispersed so the sample size upon which results are based is extremely small.

However we do have some inkling of expansion rates over time, as the expansion rates of any past time distance surface are generally consistent across the surface. Also the red shifts that are used to measure distances to the standard candles are caused not by the recession velocities of the objects emitting the light as often thought but by the cumulative stretching of space over the entire distance between us and the object. So red shifts do give us some sense of the expansion rates of the spaces and times between us and the standard candles especially when red shifts from different past times are correlated.

So an accelerating Hubble expansion is a reasonable conclusion, even though again the closer in time and space the standard candles are the less precise the expansion rates of the universe there are. In the scale of human history we have no information whatsoever because the nearest standard candle is many light years distant.

So both inflation and the varying expansion rates of space over time are completely consistent with the hyperspherical geometry of the universe when we take its radius as a P-time dimension rather than a clock time dimension. Thus the P-time radius of the universe is going to be much larger than its apparent clock time radius of 13.8 billion years, and its nearly flat observed curvature is completely consistent with that.

SEEING ALL 4 DIMENSIONS

We can actually confirm the 4-dimensional hyperspherical geometry of the universe visually because we can actually see it with our own two eyes. There has been much discussion about how to visualize the 4-dimensions of spacetime, of how to see the time dimension just as we see the 3 spatial dimensions. However the fact is we already see all 4 dimensions of the universe all the time laid out clearly before our eyes.

We see down the time dimension into the past as distance in every direction from every point in our 3-dimensional space. This is called our light cone and it's our personal view of the 4-dimensional cosmic geometry of the universe from the singularity of our location in our present moment of spacetime.

We see all 4-dimensions but there's a catch because the light cone we see is only a slice through 4-dimensions rather than the entire 4-dimensional universe. We see the past only as it existed at certain distances, and we see surfaces of space only as they existed at particular times in the past. Thus we neither see all of space nor all of time, but only a slice through both centered on our singularity.

Our experience of the passage of time through the present moment is our direct experience of the fundamental process of the universe, the continual recomputation of the information state of the universe including the passage of clock time through the present moment. This manifests as the 4-dimensional hyperspherical spacetime we directly observe around us.

SINGULARITIES IN TIME

Our location in spacetime is a singularity in the sense that clock time continuously flows into existence through the point of our location and then out in all directions into the past. Thus only our own current location exists in the present moment on our own clock. Everything else in the universe is at some distance from us and thus exists at least slightly in the past relative to us from the perspective of our present moment. Thus every observer exists alone in his own clock time singularity from which he observes the rest of the universe.

Of course everything and all observers actually exist in the same universal present moment but that common existence is not directly observable due to the finite speed of light. Our actual experience of all other things and observers in our present moment is always a no longer existent past representation down the radial time dimension of the universe.

Clock time continuously flows in from non-existence through our singularity into the present moment. The future continuously becomes the present as the state of the universe is continuously recomputed at the point of our existence. But there is no actual future that we reach that then becomes the present. The present state of the universe is just continuously recomputed in the present moment and clock time is simply the local observational rate at which those computations happen.

Though only the present moment has reality an observer can think of clock time as continuously becoming into being at his singularity and then flowing out into the past into the distance along the radial time dimension in every direction. Everywhere we look in the universe we see the past receding from us from the back of the moving train of time into the distance along the radial time dimension of the universe.

Everywhere we look we look into the past receding from our eyes, and nowhere do we see the future approaching except in our imagination. Thus our singular location in space and time is the point of the continual creation of existence, and once created the universe flows out into the past in all directions away from us.

SPACE TRAVEL

Due to the STc Principle a clock moving through space will run slower than a clock at rest, the slowing depending on the spatial velocity along its world line relative to computational space. Traveling along a world line in space will always take less time on the traveler's clock than the clock of a stay at home observer.

This slowing of a clock traveling through space can be enormous as its velocity approaches the speed of light, a fact that makes interplanetary travel theoretically feasible, at least with respect to the time required. Calculations show that a trip from earth to the center of the galaxy at a constant 1g (the equivalent of earth's gravity) acceleration for half of the trip and a 1g deceleration for the other half would take only 42 years on the clocks of the travelers, though well over 42,000 years, a little over the distance in light years to the center of the galaxy, would pass on clocks back on earth or at the galactic center (Misner, Thorne & Wheeler, 1973).

Of course a propulsion system that could produce a constant 1g acceleration for 42 years is not currently available and the difficulty of detecting and avoiding any intervening objects at close to light speed is near impossible. Nevertheless time dilation does make interstellar travel a theoretical possibility. So alien civilizations, if they exist, could just as easily travel to earth as well. The time it would take on their clocks would be quite acceptable even though it would take a very long time by our earth clocks and by clocks back on the aliens' home planet.

TIME TRAVEL

There are many misconceptions about time travel, especially when the significance of a universal present moment is not understood. However when we understand that the present moment is the only actual locus of existence and clock time runs at different rates within it everything becomes clear.

Understanding time travel and its constraints is pretty simple if we just keep in mind these two basic principles. First everything always exists in the same universal present moment and can never leave it because all that exists does so within this universal present moment and there is nowhere outside of it go. And second clock time flows at different rates within this universal present moment based on the presence of spatial velocity. That's really all there is to time travel.

We are all already traveling through time at the speed of light on our own comoving clocks all the time. So we are all already time travelers in this respect. We can't not travel in time because the passing of time is precisely us traveling through time at the speed of light. This is a basic implication of relativity. What we experience as the passage of time is us traveling in time at the speed of light.

If we are moving fast through space or are in a strong gravitational field our spatial velocity is subtracted from our velocity in time and our clocks slow down. Thus we may travel through the present moment at different clock time rates depending on our relativistic conditions, either because we have different relative motions or are in different gravitational potentials.

Though we are all traveling through clock time at different rates we all stay in the common universal current present moment. There is no possibility of traveling out of the universal present moment because it's all that exists. The present moment is the only locus of reality and of the entire actual universe.

These are the actual limits on time travel. No going back to the past or forward to the future. The future doesn't exist so there is nowhere there to go. Likewise the past doesn't exist so there is also nowhere to go in the past either. And sorry no wormholes through time either (Thorne, 1994, 483). We all stay in the common universal present moment, but our clock times, and all associated physical processes including our aging, can progress at different rates within the common present moment.

So we certainly can travel in time at different rates in the forward direction of time's arrow. There is extensive observational proof of this and relativity describes it precisely. Space traveling twins can separate and meet up again with different ages, but this is not the same as actually traveling into the future or the past by either twin. Both stay in the same universal present moment at all times and can never leave it. One just ages faster than the other in that present moment. The notion of traveling to an actual past or future out of the present is simply impossible.

So unfortunately there is no going back in time to view dinosaurs, and also no going back in time to change things there that alter the present. Thus there are no possible time travel paradoxes. And as interesting as it would be, no arrival in the present of time travelers from the future. It's simply impossible because the future has never existed because it hasn't been computed and therefore doesn't exist. It doesn't exist until it's actually computed in the present moment and then it becomes the present.

We have all arrived in the present moment from the past, but we have never left the continually evolving universal present moment to do so. But it is true that the past a space traveler arrived from could be very far back in time by our clocks if his clock was running very much slower than ours. With the right interstellar flight plan he could have left Earth when Nero was emperor of Rome and arrived back here just yesterday not a lot older than when he left Rome. In the colloquial sense that could be called a person arriving in the present from the past even though he never actually left the present moment as the centuries passed.

And neither he nor we nor anyone could ever travel back to his or any actual past since only the present moment exists and the entire past including the Roman Empire is irretrievably gone.

So we can certainly arrive at the same location in the present moment from different original past times, and that could certainly be very interesting, but everyone is continually in the common current universal present moment during the entire duration of his or her lives and travels. Some lives could be very much longer than others according to other clocks but only if they lived at much slower rates.

Our ancient Roman space traveler could arrive back on earth today to meet his 60th generation grandson and catch up on 2000 years of missed Earth history. Again extremely interesting but at every second during those 2000 years he and the earth would have both existed in the same current present moment. Events on earth would have just been progressing at a much faster clock rate than aboard the Roman space ship.

It is also theoretically possible for you or I to travel to an arbitrary date in the future in the same colloquial sense by taking a space flight with the right velocity and slowing down our clock time. But this is just a matter of slowing our clock in the universal common present moment relative to the rate of clocks at our destination. No one ever leaves the common present moment but we could arrive there with much less elapsed time on our own clock. However this requires a very high velocity space flight or intense gravitational field.

Because our spacetime is very close to flat on earth and we have a very low velocity relative to the background there is no way anyone else's time could be running appreciably *faster* than our own and there are really no additional effects to consider on that basis.

Everything that exists always exists in the same universal common present moment at all times as it evolves but time travelers could certainly arrive in the present from deep in the past with first hand information and even photos and videos given the proper technology. We can only hope!

TIME REVERSAL

There are two cases to time reversal, first time itself reversing and second an individual object or person traveling backwards in forward flowing time. We must also be careful to distinguish between the flow of time itself, and the apparently irreversible changes that occur within it such as aging and the flow of water downhill.

First there is no reason to believe that time itself can run backwards. The direction of time is due to the sequential nature of the processor cycles that compute the universe. Without a reversal in that sequence time can't flow backwards so it always flows in the same direction. But it's really a moot problem since whatever direction the processor sequence proceeds that determines the direction of time. Since the sequence of processor cycles is what determines the direction of time it's meaningless to imagine it reversing because there is no background reference with which to measure its direction since it itself is the ultimate reference. Thus by definition time always runs 'forward'.

So we can try to imagine processes running in different directions within time. But the very notion of time itself running backwards is nonsensical because whatever direction it runs is by definition the forward direction.

Some physicists have seen the apparent non-reversibility of certain processes as an unsolved problem. For example broken eggs never spontaneously reassemble into unbroken eggs. Water waves radiating from a dropped stone never reverse direction and converge to pop the stone back up out of the water, and people never reverse their aging process and start growing younger again.

These are all irreversible temporal processes yet the equations of science seem to describe them perfectly well in both temporal directions, so traditional interpretations of science offer no explanation for their irreversibility.

In Universal Reality this is a pseudo problem. First the equations that describe these processes are not actually computing them but only describing them. All such processes are the emergent manifestations of elemental computational processes that are *not time reversible* since they all involve non-reversible random choices among probability distributions at the quantum level. Once a single specific choice is made from a probability distribution of possible choices there is no way to reverse that choice into the probability distribution it was selected from. So I don't see any real problem here. It's just the way our computational universe naturally evolves as a consequence of its complete fine-tuning.

Another deeper problem of time reversibility is the nature of the time parity particle component. Elementary particles all have a number of particle components, among them space and time parity or chirality, whose nature and function are not well understood. They are clearly related to the difference between standard and anti-particles because their values are opposite in antiparticles. In fact to change particles into anti-particles and vice versa you just reverse their charges and their spatial and temporal parities (Wikipedia, Antiparticle).

Parity is the handedness of a particle. An antiparticle is like a regular particle reflected in a mirror in all three spatial dimensions, and in a time mirror as well. But what this means is not completely clear.

However 4-dimensional parity seems to be one of the elemental components of reality necessary to make something real in our universe. When particles and antiparticles meet and annihilate into energy their opposite parities cancel each other out into nothing. When particles and antiparticles emerge in pairs out of the quantum vacuum the non-actuality of the quantum vacuum can be thought of as separating into opposite actualized particle components including opposite parities. It is as if all the something in the universe other than velocity is just opposite amounts of nothing.

Thus it's reasonable to assume that dimensionality incorporates a largely normal parity as it's computed and the interaction of antiparticles with it involves the incorporation of opposite parities with respect to it. Thus spacetime itself seems to have an intrinsic normal spatial and temporal parity computed into its fabric with respect to which the parities of individual particle events are computed just as it embodies an absolute dimensional framework that linear motion and rotation are relative to.

In quantum physics, for example in their Feynman diagram representations, antiparticles must be thought of as moving backward in time to make sense of their interactions with normal particles (Wikipedia, Feynman diagram) (Feynman, 2006). However it's unlikely an antiparticle could actually be moving backwards in time. It wouldn't disappear out of the present into the past as we might assume but would be simply be seen as a particle getting younger as time progressed. However that would imply a creation date in the future which is impossible and if the particle were to annihilate and vanish at some point in time it would still continue to exist as it neared its creation date in the future.

Such contradictions can't exist in a computational universe so we have to assume that antiparticles are somehow just *pointing* backwards in time while actually going forward in time and normal particles are facing forward in time in the same direction time is flowing.

This backward facing direction in time is likely related to the problem of why there are comparatively very few antiparticles remaining in our universe when equal numbers of particles and antiparticles were presumably created together in pairs out of the quantum vacuum in the big bang. There doesn't seem to be any evidence that all those missing antiparticles are hiding anywhere.

Why this is true is uncertain but it seems like particles facing in the wrong direction in time have a much tougher time than particles facing in the direction they are going.

Note also that P-time has no meaningful time direction other than the one it creates by its own existence that defines the single possible direction of clock time's arrow. Computations just occur as happening computes them and that then creates a direction to the clock time it produces. Clock time is the effective relative rate at which processes occur in relativistic situations. It's not the rate at which P-time computes them but the rate at which their temporal aspects occur after their relative spatial motion has been computed so that the sum of both rates always equals the speed of light.

So there is no reason to believe a reverse flowing clock time is even computable by the sequential processor cycles of happening. The only way a reverse clock time could be observable would be as some physical process running backwards, but it would just be running backwards in a forward flowing clock time like a movie played backwards. Time wouldn't reverse, only the movie.

So it seems there is no possibility of clock time actually running backwards and even the concept appears meaningless. If the clock time of the entire universe ran backwards everything would appear the same because all observers' clocks and minds would be running backwards as well. And if some particular process was observably running backwards it would have to be running backwards relative to the forward flowing time of observers so clock time would still be running forwards. If there was an actual reverse direction of some local clock time, it's not clear that would be observable even if it existed.

The reverse time parity of antiparticles is an indivisible unit and single units don't actually progress in time, they can only be pointed in time. Only sequences of multiple units can be understood as having a direction in time. A single frame from a movie is equivalent to a still photo and has no direction in time. Thus an antiparticle with backward facing time parity doesn't actually move backwards in time. It's analogous to a still from a movie made to play backwards but viewed in a time moving forward. In itself one can't tell, but only when it interacts with another particle does its temporal orientation become evident.

There are other problems with individual processes moving backwards in forward flowing time. Suppose you are moving forward normally in time but at some point reverse your direction in time and begin to move backwards. First this is impossible as the current present moment in which everything exists continues to move forward and would leave you behind in a non-existent past.

But assume for a moment you could travel backwards in time. What would you see? Assuming you stepped slightly to the side to avoid a collision you would see yourself standing beside you pointing forward in time as you moved backwards. For every second retraced you'd see yourself as you were at that time getting younger reliving your life backwards.

But this raises a number of problems. First it's unclear that you would 'see' anything, as the photons from other things would now be leaving your eyes back towards them. Second it seems to imply the instantaneous creation of another you, the appearance of formed matter out of nothing, and it's not clear which of the two you's you actually would be, or are you 'now' both of you? Third would you be getting younger along with your other, or would you be getting older as he got younger? And it implies the reversibility of random quantum processes and a number of logical contradictions, which is impossible in a computational universe.

But now consider being your original self in the same situation not yet having turned around in time. Now what do you see? Now at every second you see your turned around self moving forward in time beside you, but facing backwards in time getting younger! And you should see him quite well because light is bouncing off him into your eyes normally. It's clearly an impossible scenario that can't exist because the current present moment in which everything exists is either now far advanced from you and your anti-you or was in a future outside the present moment, which is also impossible.

If for example there were some mechanism by which forward flowing time simply moved past the future creation dates of antiparticles they would vanish explaining the great preponderance of normal particles observed in the universe today but this again seems impossible.

These impossible scenarios are mentioned only because they shed light on the contradictions that would arise if a universal present moment in which everything exists didn't exist. However it is clear there is something deeper hidden in the nature of spacetime parity but for now it remains a mystery.

ENTROPY & TIME

Entropy is the tendency for the energy states in any isolated volume of space to reach equilibrium over time. For example in a completely insulated box objects at initially different temperatures will eventually all reach the same temperature. Thus presumably the entire universe will eventually reach an energy equilibrium in which no additional transfer of energy can occur and all processes will come to a halt (Wikipedia, Heat death of the universe).

This energy equilibrium is not perfect nor is it necessarily eternal due to random zero-point energy fluctuations in the quantum vacuum, which are not subject to entropy and continually affect the state of the universe of actual particles. But these effects are statistically extremely unlikely to produce any large-scale energy imbalances that affect the progression towards maximum entropy.

Because entropy appears to be a fundamental unidirectional process in time that seems irreversible some physicists have proposed that it's somehow the source of the arrow of time but this is not correct (Price, 1996, 22). We have already correctly identified the STc Principle as the source of the arrow of time, and more fundamentally the fact of the happening of existence, which continually computes the evolution of the universe, is the ultimate source of clock time and its arrow.

And entropy can't be the source of the arrow of time because it varies wildly from region to region. There are many areas of the universe in which entropy is decreasing due to incoming energy and there is certainly no reversal of the arrow of time in those areas. If entropy were responsible for the arrow of time it would have to be a universal aggregate effect rather than a local effect.

However there is no physical mechanism that could account for such a universal effect. For one thing entropy is entirely a *result* of physical processes rather than the *cause* of anything. And more importantly entropy depends entirely on the current mix of fundamental forces at any location.

Entropy states are not fundamental, as usually assumed, because they depend on the spatial mix of prevailing forces. For example cosmic scale entropy states reverse if gravitation reverses, and at smaller scales entropy depends on the distribution of the other three fundamental forces.

In an initially stable universe with only attractive gravitation the ultimate maximum entropy state will be a single black hole because all matter will eventually clump together. But in a universe with only repulsive gravitation the ultimate maximum entropy state will be a continually expanding universe in which all matter continues to fly apart forever. Thus entropy reverses if gravitation reverses.

In our expanding universe where there is an apparent mix of attractive and repulsive (dark energy) gravitation and that mix seems to be changing it's unclear what the final maximum entropy state will be.

Thus cosmological discussions of entropy are almost always

flawed because they fail to recognize that entropy itself is not fundamental (Penrose, 2005, 690). What is fundamental is the force mix including the expansion of space that defines the measure of entropy. Entropy is meaningless without reference to the force mix it's relative to. Maximum entropy has to be redefined as a state of energy equilibrium *under the mix of prevailing forces*.

Thus entropy is not a fundamental principle as usually thought. It's entirely a *result* of the evolution of the actual fundamental computational principles. Like all emergent laws it describes reality but doesn't actually compute anything.

When the dependence of entropy states on force mix and distribution is understood it becomes clear entropy has no causal connection to time and is certainly not the source of the arrow of time.

ILLUSIONS OF TIME

Our discussion of time wouldn't be complete without covering a couple of basic illusions in the way we experience it. There are many ways in which our mind's simulation of reality obscures its true nature and this is certainly true of time. The first illusion is our perception of a present moment with duration, and the second is the fact that our mind makes us think we live slightly in the future (Owen, 2016, 400).

The actual duration of the present moment is the time it takes to complete a P-time tick of happening which is the time it takes to recompute the information state of the universe. This duration is many orders of magnitude below the attosecond scale. An attosecond is equal to 10^{-18} of a second (one quintillionth of a second). For context, an attosecond is to a second what a second is to about 31.71 billion years. Thus the duration of the actual present moment is far below the resolution of human temporal perception and even far below our finest observations of quantum interactions.

Thus if our simulation accurately represented the duration of the present moment as it actually exists our entire experience would consist only of the precise current state of things in the exact infinitesimal moment. There would be no time to retain and compare before and after states of anything or the context of any event. Thus meaningful knowledge would simply be impossible. We wouldn't see any motion at all and there would be no sense of change whatsoever.

Thankfully our simulation represents the present moment with a several second duration. It holds time open just long enough that we are able to compare before and after states and observe the context of events as they occur. This is accomplished by a short-term memory routine that holds representations of events together in a sort of cache memory long enough they can be compared before tagging them as past events and moving them to long term memory if required. Without this illusion of a present moment with duration we would experience reality as inanimate objects do, completely real but completely unconsciously.

This slight opening of the present moment in time is an essential aspect of knowledge and consciousness but doesn't accurately represent the near infinitesimal duration of the actual present moment of existence. So our perception of our existence in a present moment that lasts long enough for us to make sense of things happening is a complete illusion, but an illusion essential for our existence. We can easily confirm it in operation in the visual tracks of birds and moths and by carefully examining the passage of time through consciousness.

If we rest with eyes closed and listen to relaxing music or even a single preferably low tone and progressively direct our attention closer and closer to the exact instant that it appears into and vanishes out of existence we can finally experience a state of instantaneity of time, a vanishingly short duration present moment and we suddenly realize the true nature of the present moment of time. It's a vanishingly short instant, and within that nearly non-existent moment lives the entire existence of the universe and us as well. This is an essential aspect of realizing the true nature of time as we are able to watch our short-term memory at work.

A second way our simulation of reality misrepresents time is by projecting processes slightly into the future when that's of course actually impossible since by definition the future has not actually been computed. Our minds are continually actively building a simulation of the current state of our surroundings that includes projecting current short-term processes slightly into the future and representing them to us as if they are already happening. So what we appear to see happening around us is our mind's prediction and simulation very slightly into the future of what it expects to happen.

From an evolutionary perspective this gives us an active advantage in preparing for possible future events slightly before they occur but of course these projections can't always be accurate and are continually corrected by inputs from actual events as they occur. This correction process is usually ignored by consciousness but sometimes results in a slightly shocked recognition that we saw something wrongly. With practice this can be experienced as a slight leading edge to events as they become conscious. When we observe the precise millisecond that events occur in consciousness we find they always have a fraction of a second leading edge to their occurrence.

CONCLUSION

The central experience of our existence is our consciousness in a present moment of time within which happening occurs and clock time passes at the speed of light. The present moment is universal and is simply the presence of reality. Thus all observers in the universe exist within the same universal present moment in which the entire universe exists and there is not even nothing outside, before or after.

It's clear from relativity that clock time passes at different rates depending on the presence of spatial velocity within this shared universal present moment. It is also clear from relativity that all observers in the universe continually travel forward in clock time at the speed of light as measured by their own clocks. And it's clear that all observers see the 4th dimension of past clock time as distance in every direction from their location in the 3-dimensional space of the universe.

All these aspects of time can be directly realized in our experience. If we turn our attention to the passage of happening and clock time through the present moment we find our consciousness of this process is indeed the fundamental experience of our existence. We just need to realize that this experience is us and everything around us traveling at the speed of light through the 4th dimension of time even while we sit on our sofas. We are all surfing the 3-dimensional surface of our expanding hypersphere at the speed of light as we ride the evolving wave of existence.

And with the assistance of science we can directly experience the fact that clock times passes at different rates depending on the presence of spatial velocity. If we observe the half-lives of decaying particles moving at relativistic rates, the speed of our clocks on earth relative to those traveling in space, or even by directly comparing our clocks to those returning from space flights we can directly experience this. They can all be directly realized in relativistic circumstances in our daily lives. Even magnetism is our direct experience of the relativistic effect of electric charges moving with relativistic velocities (Owen, 2016).

We can also directly experience and realize the continual computational creation of the information state of the present as a process that occurs only within the happening of the present moment, thus realizing the non-existence of the future. We can also directly realize the non-existence of the past even though we observe it as distance in every direction because we are observing that and everything else in the universal present moment of all existence as its light arrives in our eyes.

Thus we immediately realize the impossibility of time travel in the sense of traveling out of the present moment. The present moment is all that exists and where everything exists and happens. We can see down the past dimension of time only because of the finite speed of light. We are not actually observing the past, but the light trace of the past in the present moment.

The very fabric of space itself consists of the fixed speed of light velocity of spacetime at every point. That velocity can be either velocity in space or velocity in time but is always equal to the speed of light c. Thus whenever velocity in space increases velocity in time decreases and this STc Principle is the key to understanding both time and relativity.

The two fundamental aspects of time are the present moment and happening. The present moment is simply the manifestation of the presence of reality. And happening is the continual recomputation of the data that constitutes the universe.

The rate of happening is what we call clock time. If we think of the universe as a computational system the processor that computes the universe has a fixed cycle rate that is the direct source of the fixed spacetime speed of light velocity of every point and process in the universe. When velocity in space increases there are fewer processor cycles left over to compute velocity in time and time dilation occurs. This is the computational source of the STc Principle and most of the effects of relativity.

Putting thus all together we arrive at a deep and comprehensive understanding of the mystery of time, what it is and how it works, and its major implications for the nature of reality itself.

NOTES

1. DERIVING RELATIVITY FROM THE STc Principle

For the mathematically inclined the Lorentz transformation that governs most of special relativity can be derived from the STc Principle as follows indicating the STc Principle underlies special relativity. To simplify the discussion we assume that any spatial velocity is parallel to the x-axis. Then the STc Principle can be expressed mathematically as

$$\mathbf{v}_x + \mathbf{v}_T = \mathbf{c} \tag{1.1}$$

Where $\mathbf{v}_x$ is the velocity through space along the x-axis, $\mathbf{v}_T$ is the velocity through time, and $\mathbf{c}$ is the velocity of light. Writing the quantities in bold indicates that they are vectors; that is they have both a magnitude and a direction. Note that velocity in time is actually the relative time rate of any clock to an observer's clock multiplied by the speed of light c to put it in the same units as spatial velocity so time can be treated as another dimension of a consistent 4-dimensional geometry.

Expressing the x velocity as a vector $\mathbf{v}_x$ is standard physics, however some physicists might recoil at expressing the velocity of time as a vector since it's normally considered to be a scalar, a quantity having a magnitude but no direction. In fact though, in a 4-dimensional universe, time clearly does have a direction along the time axis, and thus is most certainly a vector. The addition of two vectors always produces another vector, so that the result of Eq. (1.1) is a vector *velocity* in a particular direction and the magnitude of that velocity is always the scalar *speed* of light.

Eq. (1.1) can be depicted graphically. In the following figure the vertical axis is the velocity of time, and the horizontal axis is the velocity in space along the x-axis. When we plot Eq. (1.1), we get a circle of constant radius c = the speed of light. The graph shows an arbitrary example whose spacetime velocity vector extends from the origin to the circle along with its space and time velocity components projected on their respective axes. It also shows the cases in which the c spacetime velocity is directed either entirely along the vertical time or horizontal space axis.

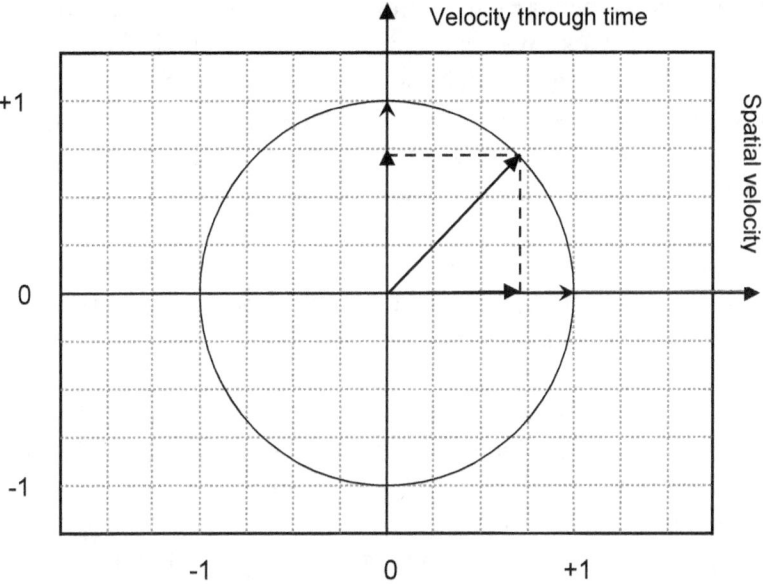

Fig. 1.1. The STc Principle. The vector sum of space and time velocities of every entity in the universe is always equal in magnitude to the speed of light c at every moment, as shown by the circle. Arrows show the three possible cases. When there is no relative velocity in space all an object's spacetime velocity is through time at the speed of light as shown by the vertical arrow touching the circle. In contrast, light itself always travels with relative velocity in space $v_x = c$ as represented by the horizontal arrow touching the circle. In the third case, when $0 < v_x < c$ the spacetime velocity is the vector sum of the projections of the space and time velocities along their respective axes as shown by the diagonal arrow. But in all cases the vector sum of space and time velocities is equal in magnitude to the speed of light as indicated by the circle. Though the c velocity is represented as a complete circle in the figure only velocities in the upper half circle where time is positive are actually possible.

The mathematical rules of vector addition are just expressions of simple geometry and the Pythagorean theorem so that we can alternately express Eq. (1.1) as

$$v_x^2 + v_T^2 = c^2 \qquad (1.2)$$

That is the sum of the squares of the space and time velocities is equal to the square of the speed of light. This is the more usual mathematical form equivalent to the vector form of (1.1).

From Newtonian physics we know that a velocity is the time rate of change of a distance so that the space velocity along the x axis could also have been written as dx/dt, the instantaneous rate of change of distance x with time t at any given moment.

Now what is meant by the velocity of time? The idea is that a moving clock runs at a slower rate than a stationary clock. So it makes sense to consider the rate at which the moving clock runs relative to an observer's stationary clock as the relative velocity of the time displayed by the moving clock. We express this velocity as dT/dt, the time rate of change of the moving clock as measured by the stationary clock.

Another consideration is that distance in space is measured in spatial units such as meters, whereas distance along the time axis is measured in time units such as seconds. To have a meaningful equation in which both types of distances and their velocities can be compared they must both be expressed in the same units. This can be done by expressing both in a unit such as light seconds, the distance that light travels in a second, $\approx 3 \times 10^8$ meters. This is easily accomplished by substituting cT for T to express the distance along the time axis measured by the moving clock. This gives the equivalent time distance (seconds) in units of spatial distance (light seconds). With these additions we can now rewrite Eq. (1.2) as

$$v_x^2 + \left(\frac{dcT}{dt}\right)^2 = c^2$$

and since for a constant c, dcT/dt = cdT/dt

$$v_x^2 + c^2\left(\frac{dT}{dt}\right)^2 = c^2 \qquad (1.3)$$

Rearranging and dividing both sides by c^2

$$\left(\frac{dT}{dt}\right)^2 = \frac{c^2 - v_x^2}{c^2} = 1 - \frac{v_x^2}{c^2}$$

and taking the square root of both sides we have

$$\frac{dT}{dt} = \sqrt{1 - \frac{v_x^2}{c^2}} \qquad (1.4)$$

which gives the velocity of time (the relative slowing of its clock rate) for a clock moving at velocity v_x relative to the observer's clock.

Now Eq. (1.4) is a fundamental equation of special relativity, an expression of the Lorentz transform, and from it all the standard effects of special relativity can be derived except for $E = mc^2$ which follows with the addition of the standard classical principles of conservation of energy and momentum. Thus we find that the theory of special relativity is in fact a consequence of the STc Principle.

2. THE BLOCK TIME DELUSION

Some physicists deny the concept of a present moment and instead believe in the block time theory in which all times exist at once and there is no preferred present moment (Wikipedia, Eternalism (philosophy of time)). This is a fundamental error based on a mistaken interpretation of relativity.

Relativity clearly reveals the universe as a 4-dimensional structure but that doesn't require the time dimension to have any actual extension. All we need to satisfy relativity is a 4-dimensional surface where the only moment in time that exists is the current universal present moment in which clock times are being computed.

The standard block universe model envisions the total past present and future history of the universe existing as a single fixed completely static 4-dimensional structure in some sort of vaguely defined eternal time or outside of time. Our apparent existence in this particular present moment we seem to experience is an illusion because we are also continually having the same illusory experience in every point of time of our block time world line corresponding to every static instant of our life.

In a block time universe the entire history of the universe already exists. The past still exists and the future is already fixed and exists as well. Thus it's completely fixed and deterministic and can never be any different than it is or changed it the least because change doesn't exist.

There are so many things wrong with this theory it's hard to know where to begin but we will mention the main ones.

1. It's contradicted by relativity itself since the STc Principle clearly demonstrates that everything in the universe has velocity in time, which necessitates it being at one privileged location in time at any moment and that is where the present moment is. Thus

contrary to the beliefs of some physicists relativity doesn't even work without a present moment.
2. In a block time universe the entire history of the universe already exists. The past still exists and the future is already fixed. Thus it's completely deterministic and fixed and can never be any different than it is. There is no randomness, chance or freedom whatsoever. Life is essentially meaningless.
3. In a block universe we are simply a sequence of static slices of our time line like stills from a movie and each of those static slices is having the illusion it's in a present moment. But if nothing is ever actually changing there is no way whatsoever that what is effectively a still photo of ourselves could be having any experiences at all. In short consciousness itself is impossible in a block universe because the experience of consciousness is an experience of something happening.
4. Block universe advocates tell us that the us and them that seem to be conversing right now are just static time slices of us both but there is no explanation whatsoever for why it's these particular static time slices that have been selected to do the talking. Somehow all our static time slices are always doing their thing but there is no attempt to explain why 'we' are the slice that we actually experience, nor is there any attempt to explain the obvious temporal progression from slice to slice our experience follows.
5. There is also the fundamental problem of how a block universe could come into being. It obviously has an enormously complex and consistent causal structure. How could such a structure be non-sequentially created *ab initio*? It seems to suggest the necessity of an omniscient creator god that stands outside of time, which just begs the question of how he was created. How would whatever process created it know how to create each state so that it would be completely causally consistent with the preceding one if the previous state didn't actually cause the successor state? But if previous states were the source of successor states then all states had to have been created sequentially and that is precisely the standard time evolving universe that the block universe denies!
6. A block time universe violates practically every scientific law known. For example it violates the conservation of mass-energy since the mass-energy of all time slices exists at once so the total mass-energy of a block universe is the sum across all time slices and the number of time slices could vary depending on the universe. But if we claim the conservation of mass-energy applies only to single time slices then why and why is it consistent from slice to slice?

7. It's also completely impossible to speak meaningfully about science and the universe from the perspective of belief in a block universe. No matter how hard they try even the firmest believers must immediately revert to describing any aspect of reality or daily life in terms of clock time flowing through a present moment. Thus reality just doesn't make sense from the perspective of block time and this lesson should be taken to heart.
8. Finally advocates of the block universe theory tend to be true believers and take it as a tenant of quasi-religious belief. Any logical or scientific arguments raised against it tend to be summary dismissed as attacks on tenants of faith with no cogent reason given to reject them other than it just isn't so and doesn't really work that way.

In summary the block time universe theory is total nonsense that only a physicist could have invented and must be summarily rejected as the very obvious existence of an actual present moment and the flow of clock time completely falsify it.

BIBLIOGRAPHY

Feynman, Richard. *Lectures On Physics*. Pearson, Addison, Wesley, 2006.
Greene, Brian. *The Elegant Universe*. Norton, 1999.
Greene, Brian. *The Fabric of The Cosmos*. Vintage Books, 2005.
Hawking, Stephen W. *A Brief History of Time*. Bantam Books. 1998.
Misner, Charles W.; Thorne, Kip S.; Wheeler, Archibald. *Gravitation*. Freeman, 1973.
Owen, Edgar L. *Spacetime and Consciousness*. EdgarLOwen.info. 2007.
Owen, Edgar L. *Mind and Reality*. EdgarLOwen.info. 2009.
Owen, Edgar L. *Reality*. Amazon.com. 2013.
Owen, Edgar L. *Universal Reality*. Amazon.com. 2016.
Penrose, Roger. *The Road to Reality*. Knopf, 2005.
Price, Huw. *Time's Arrow and Archimedes' Point*. Oxford, 1996.
Thorne, Kip S. *Black Holes and Time Warps*. Norton, 1994.
Wikipedia contributors. *Wikipedia, the Free Encyclopedia*. http://wikipedia.org
Wise, Alex. *Event horizon pileup*. https://wiki.physics.udel.edu/AAP/Event_horizon_pileup

Edgar L. Owen was born April 1st, 1941 and quickly realized that reality is not as it appears to be. A child prodigy, he entered the University of Tulsa aged 15 and received a B.S. with honors in science and mathematics with a minor in philosophy at 18 before completing several more years of graduate study in physics and philosophy.

In the early 60's he moved to the Haight-Ashbury in San Francisco where he hung out with notables from the Beat Generation and conducted an intense personal study of the nature of mind and consciousness. From there he traveled to Japan where he lived for three years studying Zen and Buddhist philosophy while subsisting as a ronin English teacher.

Upon returning to the US he began a career in computer science writing numerous programs in artificial intelligence, simulations, graphics, and cellular automata while designing and managing advanced computer systems for the New York Federal Reserve Bank and AT&T. He then left the corporate world to start his own software business marketing his own CAD programs, which he ran for a number of years. Currently he owns a premier Internet gallery of fine Ancient Art and Classical Numismatics at EdgarLOwen.com.

Deeply immersed in nature since childhood, and always considering it the ultimate source of his inspiration and knowledge of reality, he has served as Chairman of his local Environmental Commission and organized several campaigns to protect the local environment and its wildlife.

Over the last several years he has worked to combine and organize the results of a lifetime of study of the various aspects of reality into a single coherent Theory of Everything. He now spends most of his time exploring the wonderful awesome mystery of reality and how it can be experienced more fully and deeply and enjoying his existence within it.

Edgar currently lives in Northern NJ in a big old house on top of a hill where he communes with nature and enjoys the company of his wild visitors including the occasional human. Edgar is currently single and looking for a younger housekeeper companion ☺. He can be contacted at Edgar@EdgarLOwen.com.

www.ingramcontent.com/pod-product-compliance
Lightning Source LLC
Chambersburg PA
CBHW080545190526
45169CB00007B/2649